Compiled by Chinese Academy of Tropical Agricultural Sciences (CATAS) and
Chinese Society for Tropical Crops (CSTC)
A Series of Books for Field Guide to Common Plants in FSM

General Editor: Liu Guodao

Field Guide to Flowers and Ornamental Plants in the Federated States of Micronesia

Editors in Chief: Yang Guangsui Shen Zhen Liu Shaoshan

China Agricultural Science and Technology Press

图书在版编目（CIP）数据

密克罗尼西亚联邦花卉植物图鉴 = Field Guide to Flowers and Ornamental Plants in the Federated States of Micronesia / 杨光穗，谌振，刘少姗主编 .—北京：中国农业科学技术出版社，2021.5

（密克罗尼西亚常见植物图鉴系列丛书 / 刘国道主编）

ISBN 978-7-5116-5287-4

Ⅰ.①密… Ⅱ.①杨…②谌…③刘… Ⅲ.①花卉—种质资源—密克罗尼西亚联邦—图集 Ⅳ.① S68-64

中国版本图书馆 CIP 数据核字（2021）第 068610 号

责任编辑	徐定娜
责任校对	贾海霞
责任印制	姜义伟　王思文
出 版 者	中国农业科学技术出版社 北京市中关村南大街 12 号　邮编：100081
电　　话	（010）82105169（编辑室）　（010）82109702（发行部） （010）82109709（读者服务部）
传　　真	（010）82109707
网　　址	http://www.castp.cn
发　　行	各地新华书店
印 刷 者	北京科信印刷有限公司
开　　本	787 mm×1 092 mm　1/16
印　　张	9.25
字　　数	406 千字
版　　次	2021 年 5 月第 1 版　2021 年 5 月第 1 次印刷
定　　价	108.00 元

版权所有·侵权必究

About the Author

Dr. Liu Guodao, born in June 1963 in Tengchong City, Yunnan province, is the incumbent Vice President of Chinese Academy of Tropical Agricultural Sciences (CATAS). Being a professor and PhD tutor, he also serves as the Director-General of the China-Republic of the Congo Agricultural Technology Demonstration Center.

In 2007, he was granted with his PhD degree from the South China University of Tropical Agriculture, majoring in Crop Cultivation and Farming.

Apart from focusing on the work of CATAS, he also acts as a tutor of PhD candidates at Hainan University, Member of the Steering Committee of the FAO Tropical Agriculture Platform (TAP), Council Member of the International Rubber Research and Development Board (IRRDB), Chairman of the Chinese Society for Tropical Crops, Chairman of the Botanical Society of Hainan, Executive Director of the Chinese Grassland Society and Deputy Director of the National Committee for the Examination and Approval of Forage Varieties and the National Committee for the Examination and Approval of Tropical Crop Varieties.

He has long been engaged in the research of tropical forage. He has presided over 30 national, provincial and ministerial-level projects: namely the "National Project on Key Basic Research (973 Program)" and international cooperation projects of the Ministry of Science and Technology, projects of the National Natural Science Foundation of China, projects of the International Center for Tropical Agriculture in Colombia and a bunch of projects sponsored by the Ministry of Agriculture and Rural Affairs (MARA) including the Talent Support Project, the "948" Program and the Infrastructure Project and Special

Scientific Research Projects of Public Welfare Industry.

He has published more than 300 monographs in domestic and international journals such as "New Phytologist" "Journal of Experimental Botany" "The Rangeland Journal" "Acta Prataculturae Sinica" "Acta Agrestia Sinica" "Chinese Journal of Tropical Crops", among which there nearly 20 were being included in the SCI database. Besides, he has compiled over 10 monographs, encompassing "Poaceae Plants in Hainan" "Cyperaceae Plants in Hainan" "Forage Plants in Hainan" "Germplasm Resources of Tropical Crops" "Germplasm Resources of Tropical Forage Plants" "Seeds of Tropical Forage Plants" "Chinese Tropical Forage Plant Resources". As the chief editor, he came out a textbook-*Tropical Forage Cultivation*, and two series of books-*Practical Techniques for Animal Husbandry in South China Agricultural Regions* (19 volumes) and *Practical Techniques for Chinese Tropical Agriculture "Going Global"* (16 volumes).

He has won more than 20 provincial-level and ministerial-level science and technology awards. They are the Team Award, the Popular Science Award and the First Prize of the MARA China Agricultural Science and Technology Award, the Special Prize of Hainan Natural Science Award, the First Prize of the Hainan Science and Technology Progress Award and the First Prize of Hainan Science and Technology Achievement Transformation Award.

He developed 23 new forage varieties including Reyan No. 4 King grass. He was granted with 6 patents of invention and 10 utility models by national patent authorities. He is an Outstanding Contributor in Hainan province and a Special Government Allowance Expert of the State Council.

Below are the awards he has won over the years: in 2020, "the Ho Leung Ho Lee Foundation Award for Science and Technology Innovation"; in 2018, "the High-Level Talent of Hainan province" "the Outstanding Talent of Hainan province" "the Hainan Science and Technology Figure"; in 2015, Team Award of "the China Agricultural Science and Technology Award" by the Ministry of Agriculture; in 2012, "the National Outstanding Agricultural Talents Prize" awarded by the Ministry of Agriculture and as team leader of the team award: "the Ministry of Agriculture Innovation Team" (focusing on the research of Tropical forage germplasm innovation and utilization); in 2010, the first-level candidate of the "515 Talent Project" in Hainan province; in 2005, "the Outstanding Talent of Hainan province"; in 2004, the first group of national-level candidates for the "New Century Talents Project" "the 4th Hainan Youth Science and Technology Award" "the 4th Hainan Youth May 4th Medal" "the 8th China Youth Science and Technology Award" "the Hainan Provincial International Science and Technology Cooperation Contribution Award"; in 2003, "a Cross-Century Outstanding Talent" awarded by the Ministry of Education; In 2001, "the 7th China Youth Science and Technology Award" of Chinese Association of Agricultural Science Societies, "the National Advanced Worker of Agricultural Science and Technology"; in 1993, "the Award for Talents with Outstanding Contributions after Returning to China" by the State Administration of Foreign Experts Affairs.

Dr. Yang Guangsui, born in September 1972 in Kaili City, Guizhou province, is a Professor in the Tropical Crops Genetic Resources Institute of Chinese Academy of Tropical Agricultural Sciences. She is mainly engaged in the innovation of tropical flower germplasm resources, seedling breeding and flower cultivation technology for industrialization. In the past 3 years, she has established 3 germplasm nurseries of tropical waterlily, Bougainvillea and zingiberaceae flowering plants, collected more than 3,700 resources and over 1,536 species, evaluated a total of 644 resources. The resource evaluation system for waterlily, hippeastrum rutilum and zingiberaceae flowering plants has been developed, the breeding system for waterlily and hippeastrum rutilum has been set up and the seedling propagation technology system for Bougainville, waterlily and hippeastrum rutilum has taken shape. She has presided and participated in more than 10 provincial-level and ministerial-level research projects of the Ministry of Agriculture and Rural Affairs and Hainan province and won one Second Prize and one Third Prize of the Hainan Science and Technology Progress Award. Seven varieties developed by her were approved in Hainan province. She was granted with one plant right and three patents. Over the years, she has published 19 papers (including 3 SCI papers), 3 monographs as Chief Editor or Deputy Chief Editor, and 3 plans on flower and leisure industry development. Moreover, she enacted and issued 2 local standards. In 2018, she was selected as the "Leading Talent" in Hainan province. Apart from fulfilling the major work at CATAS, she also acts as the master's tutor at Hainan University, member of the Hainan Variety Approval Committee of Main Forest Trees, Director of Tropical Flower Research Center of TCGRI-CATAS, Secretary General of Tropical Flower Industry Technology Innovation Strategy Alliance and the Hainan Bougainvillea Association, Deputy Secretary General of Hainan Flower Association, and Member of the Expert 110 in Hainan Agricultural Science and Technology.

A Series of Books for Field Guide to Common Plants in FSM

General Editor: Liu Guodao

Field Guide to Flowers and Ornamental Plants in the Federated States of Micronesia Editorial Board

Editors in chief:

Yang Guangsui   Shen Zhen   Liu Shaoshan

Associate editors in chief:

Wang Xiaofang   Huang Mingzhong   Wang Cun

Members (in alphabet order of surname):

Fan Haikuo	Gong Shufang	Hao Chaoyun	Huang Guixiu
Huang Mingzhong	Li Weiming	Liu Guodao	Liu Shaoshan
Shen Zhen	Tang Qinghua	Wang Cun	Wang Jinhui
Wang Qinglong	Wang Xiaofang	Wang Yuanyuan	Yang Guangsui
Yang Hubiao	Zheng Xiaowei		

Photographers:

Wang Qinglong   Yang Hubiao   Yang Guangsui

Tanslator:

Wu Minyu

The President
Palikir, Pohnpei
Federated States of Micronesia

FOREWORD

It is with great pleasure that I present this publication, "Agriculture Guideline Booklet" to the people of the Federated States of Micronesia (FSM).

The Agriculture Guideline Booklet is intended to strengthen the FSM Agriculture Sector by providing farmers and families the latest information that can be used by all in our communities to practice sound agricultural practices and to support and strengthen our local, state and national policies in food security. I am confident that the comprehensive notes, tools and data provided in the guideline booklets will be of great value to our economic development sector.

Special Appreciation is extended to the Government of the People's Republic of China, mostly the Chinese Academy of Tropical Agricultural Sciences (CATAS) for assisting the Government of the FSM especially our sisters' island states in publishing books for agricultural production. Your generous assistance in providing the practical farming techniques in agriculture will make the people of the FSM more agriculturally productive.

I would also like to thank our key staff of the National Government, Department of Resources and Development, the states' agriculture and forestry divisions and all relevant partners and stakeholders for their kind assistance and support extended to the team of Scientists and experts from CATAS during their extensive visit and work done in the FSM in 2018.

We look forward to a mutually beneficial partnership.

Sincerely,

David W. Panuelo
President

Preface

Claiming waters of over 3,000 square kilometers, the vast area where Pacific island countries nestle is home to more than 10,000 islands. Its location at the intersection of the east-west and north-south main traffic artery of Pacific wins itself geo-strategic significance. There are rich natural resources such as agricultural and mineral resources, oil and gas here. The relationship between the Federated States of Micronesia (hereinafter referred to as FSM) and China ushered in a new era in 2014 when Xi Jinping, President of China, and the leader of FSM decided to establish a strategic partnership on the basis of mutual respect and common development. Mr Christian, President of FSM, took a successful visit to China in March 2017 during which a consensus had been reached between the leaders that the traditional relationship should be deepened and pragmatic cooperation (especially in agriculture) be strengthened. This visit pointed out the direction for the development of relationship between the two countries. In November 2018, President Xi visited Papua New Guinea and in a collective meeting met 8 leaders of Pacific Island countries (with whom China has established diplomatic relation). China elevated the relationship between the countries into a comprehensive and strategic one on the basis of mutual respect and common development, a sign foreseeing a brand new prospect of cooperation.

The government of China launched a project aimed at assisting FSM in setting up demonstration farms in 1998. Until now, China has completed 10 agricultural technology cooperation projects. To answer the request of the government of FSM, Chinese Academy of Tropical Agricultural Sciences (hereinafter referred to as CATAS), directly affiliated with the

Ministry of Agriculture and Rural Affairs of China, was elected by the government of China to carry out training courses on agricultural technology in FSM during 2017—2018. The fruitful outcome is an output of training 125 agricultural backbone technicians and a series of popular science books which are entitled "Field Guide to Forages in the Federated States of Micronesia" "Field Guide to Flowers and Ornamental Plants in the Federated States of Micronesia" "Field Guide to Medicinal Plants in the Federated States of Micronesia" "Field Guide to Fruits and Vegetables in the Federated States of Micronesia" "Coconut Germplasm Resources in the Federated States of Micronesia" and "Field Guide to Plant Diseases, Insect Pests and Weeds in the Federated States of Micronesia".

In these books, 492 accessions of germplasm resources such as coconut, fruits, vegetables, flowers, forages, medical plants, and pests and weeds are systematically elaborated with profuse inclusion of pictures. They are rare and precious references to the agricultural resources in FSM, as well as a heart-winning project among China's aids to FSM.

Upon the notable moment of China-Pacific Island Countries Agriculture Ministers Meeting, I would like to send my sincere respect and congratulation to the experts of CATAS and friends from FSM who have contributed remarkably to the production of these books. I am firmly convinced that the exchange between the two countries on agriculture, culture and education will be much closer under the background of the publication of these books and Nadi Declaration of China-Pacific Island Countries Agriculture Ministers Meeting, and that more fruitful results will come about. I also believe that the team of experts in tropical agriculture mainly from the CATAS will make a greater contribution to closer connection in agricultural development strategies and plans between China and FSM, and closer cooperation in exchanges and capacity-building of agriculture staffs, in agricultural science and technology for the development of agriculture of both countries, in agricultural investment and trade, in facilitating FSM to expand industry chain and value chain of agriculture, etc.

Qu Dongyu
Director General
Food and Agriculture Organization of the United Nations
July 23, 2019

Located in the northern and central Pacific region, the Federated States of Micronesia (FSM) is an important hub connecting Asia and America. Micronesia has a large sea area, rich marine resources, good ecological environment, and unique traditional culture.

In the past 30 years since the establishment of diplomatic relations between China and FSM, cooperation in diverse fields at various levels has been further developed. Since the 18th National Congress of the Communist Party of China, under the guidance of Xi Jinping's thoughts on diplomacy, China has adhered to the fine diplomatic tradition of treating all countries as equals, adhered to the principle of upholding justice while pursuing shared interests and the principle of sincerity, real results, affinity, and good faith, and made historic achievements in the development of P.R. China-FSM relations.

The Chinese government attaches great importance to P.R. China-FSM relations and always sees FSM as a good friend and a good partner in the Pacific island region. In 2014, President Xi Jinping and the leader of the FSM made the decision to build a strategic partnership featuring mutual respect and common development, opening a new chapter of P.R. China-FSM relations. In 2017, FSM President Peter Christian made a successful visit to China. President Xi Jinping and President Christian reached broad consensuses on deepening the traditional friendship between the two countries and expanding practical cooperation between the two sides, and thus further promoted P.R. China-FSM relations. In 2018, Chinese President Xi Jinping and Micronesian President Peter Christian had a successful meeting again in PNG and made significant achievements, deciding to upgrade P.R. China-FSM relations to a new stage of Comprehensive Strategic Partnership, thus charting the course for future long-term development of P.R. China-FSM relations.

In 1998, the Chinese government implemented the P.R. China-FSM demonstration farm project in FSM. Ten agricultural technology cooperation projects have been completed, which has become the "golden signboard" for China's aid to FSM. From 2017 to 2018, the Chinese Academy of Tropical Agricultural Sciences (CATAS), directly affiliated with the Ministry of Agriculture and Rural Affairs, conducted a month-long technical training on pest control of coconut trees in FSM at the request of the Government of FSM. 125 agricultural managers, technical personnel and growers were trained in Yap, Chuuk, Kosrae and Pohnpei, and the biological control technology demonstration of the major dangerous pest, Coconut Leaf Beetle, was carried out. At the same time, the experts took advantage of the spare time of the training course and spared no effort to carry out the preliminary evaluation of the investigation and utilization of agricultural resources, such as coconut, areca nut, fruit tree, flower, forage, medicinal plant, melon and vegetable, crop disease, insect pest and weed diseases, in the field in conjunction with Department of Resources and Development of FSM and the vast number of trainees, organized and compiled a series of popular science books, such as "Field Guide to Forages in the Federated States of Micronesia" "Field Guide to Flowers and Ornamental Plants in the Federated States of Micronesia" "Field Guide to Medicinal Plants in the Federated States of Micronesia" "Field Guide to Fruits and Vegetables in the Federated States of Micronesia" "Coconut Germplasm Resources in the Federated States of Micronesia" and "Field Guide to Plant Diseases, Insect Pests and Weeds in the Federated States of Micronesia".

The book introduces 37 kinds of coconut germplasm resources, 60 kinds of fruits and vegetables, 91 kinds of angiosperm flowers as well as 13 kinds of ornamental pteridophytes, 100 kinds of forage plants, 117 kinds of medicinal plants, 74 kinds of crop diseases, pests and weed diseases, in an easy-to-understand manner. It is a rare agricultural resource illustration in FSM. This series of books is not only suitable for the scientific and educational workers of FSM, but also it is a valuable reference book for industry managers, students, growers and all other people who are interested in the agricultural resources of FSM.

This series is of great significance for it is published on the occasion of the 30th anniversary of the establishment of diplomatic relations between the People's Republic of

China and FSM. Here, I would like to pay tribute to the experts from CATAS and the friends in FSM who have made outstanding contributions to this series of books. I congratulate and thank all the participants in this series for their hard and excellent work. I firmly believe that based on this series of books, the agricultural and cultural exchanges between China and FSM will get closer with each passing day, and better results will be achieved more quickly. At the same time, I firmly believe that the Chinese Tropical Agricultural Research Team, with CATAS as its main force, will bring new vigour and make new contributions to promoting the in-depth development of the strategic partnership between the People's Republic of China and the Federated States of Micronesia, strengthening solidarity and cooperation between P.R. China and the developing countries, and the P.R. China-FSM joint pursuit of the Belt and Road initiative and building a community with a shared future for the humanity.

Ambassador Extraordinary & Plenipotentiary of
the People's Republic of China to
the Federated States of Micronesia
May 23, 2019

Foreword

    The Federated States of Micronesia is located in the central part of Pacific Ocean. It consists of 607 islands ranging from volcanic islands of mountainous regions to coral islands with beautiful sceneries. It features tropical marine climate, and rain season falls from April to November with an annual precipitation of 2,000 mm as well as an average temperature of 27°C, which makes it possible to cultivate a variety of tropical ornamental plants here.

    We were greeted with the picturesque scenery, magnificent dancing, vibrant garlands and sincere smiles when 12 of us visited this attractive place. Micronesians are indigenous inhabitants who moved here about 4000 years ago. Then, they began to develop and inherit a unique folk custom. Native people are fond of flowers very much. According to local custom of their tribe, females tend to wear skirts woven by flowers, garlands made of flowers and nypa fruticans' leaves. They look like fairies favored by Mother Nature.

    The locals love dressing themselves up with flowers, they also take delight in decorating their homes with ornamental plants. Many kinds of flowers can be seen on the roadside, in the park or in front of government buildings. Meanwhile, flowers in full bloom invigorate the whole community.

    Their preference to flowers embodies the tribe's traditional value and the profound influence of floral appreciation from other countries. They are inclined to import different kinds of tropical flowers of vibrant colors and longer florescence from America, Thailand,

Japan or China. But, it still takes a certain period of time to do a further research on the use of their native flowers such as *Ixora casei* and *dendrobium* sp.

Moon in river resembles a silver mirror; Mirage looms behind the capricious clouds. Federated States of Micronesia is not only a string of islands on the Pacific Ocean, but a romantic floral paradise.

General Editor

Vice President of Chinese Academy of Tropical Agricultural Sciences

March 22, 2019

Contents

Angiospermae

- **Labiatae** 3
  - *Plectranthus scutellarioides* 3

- **Euphorbiaceae** 4
  - *Euphorbia milii* 4
  - *Jatropha integerrima* 6
  - *Euphorbia cyathophora* 7
  - *Acalypha hispida* 8
  - *Codiaeum variegatum* 9
  - *Acalypha wikesiana* 11

- **Leguminosae** 13
  - *Caesalpinia pulcherrima* 13
  - *Calliandra surinamensis* 15
  - *Delonix regia* 17

- **Balsaminaceae** 18
  - *Impatiens balsamina* 18

- **Apocynaceae** 19
  - *Plumeria rubra* L. cv. *Acutifolia* 19
  - *Rubra forma acutifolia* cv. 'Gold' 20
  - *Plumeria rubra* cv. *Acutifolia* 21
  - *Plumeria pudica* 22
  - *Nerium indicum* 23

  - *Thevetia peruviana* 24
  - *Catharanthus roseus* 25
  - *Allemanda neriifolia* 26
  - *Allamanda violacea* 27
  - *Cryptostegia* sp. 28

- **Zingiberaceae** 29
  - *Curcuma longa* 29
  - *Curcuma australasica* 30
  - *Zingiber zerumbet* 31
  - *Alpinia purpurata* 32
  - *Hedychium coronarium* 33
  - *Hedychium coronarium* 34
  - *Costus spciousus* (J.Koenig) Sm. 35

- **Malpighiaceae** 36
  - *Tristellateia australasiae* 36

- **Malvaceae** 37
  - *Hibiscus rosa - sinensis* 37
  - *Hibiscus mutabilis* 39

- **Campanulaceae** 40
  - *Isotoma axillaris* 40

- **Asteraceae** 41
  - *Zinnia elegans* 41

- **Taccaceae** ... 42
    - *Tacca leontopetaloides* ... 42

- **Acanthaceae** ... 43
    - *Barleria cristata* ... 43
    - *Pseuderanthemum reticulatum* ... 44
    - *Pseuderanthemum reticulatum* ... 45
    - *Thunbergia grandiflora* ... 46
    - *Staurogyne concinnula* ... 47

- **Gesneriaceae** ... 48
    - *Alloplectus martius* ... 48
    - *Episcia cupreata* ... 49

- **Orchidaceae** ... 50
    - *Taeniophyllum* sp. ... 50
    - *Rodriguezia* sp. ... 51
    - *Epidendrum radicans* ... 52
    - *Aranda* Hybrid ... 53
    - *Arachnis* Maggie Oei ... 54
    - *Papilionanthe* Miss Joaquim ... 55
    - *Spathoglottis plicata* ... 56
    - *Spathoglottis micronesia* ... 57
    - *Dendrobium* sp. ... 58

- **Polygonaceae** ... 59
    - *Antigonon leptopus* ... 59

- **Agavaceae** ... 60
    - *Cordyline fruticosa* ... 60

- **Musaceae** ... 62
    - *Heliconia* spp. ... 62

- **Verbenaceae** ... 63
    - *Lantana camara* ... 63
    - *Duranta repens* ... 65
    - *Clerodendrum thomsonae* ... 67
    - *Clerodendrum japonicum* ... 68
    - *Holmskioldia sanguinea* ... 70

- **Cannaceae** ... 71
    - *Canna indica* ... 71

- **Oleaceae** ... 72
    - *Jasminum grandiflorum* ... 72

- **Vitaceae** ... 73
    - *Leea indica* ... 73

- **Lythraceae** ... 74
    - *Lagerstroemia micrantha* ... 74
    - *Lagerstroemia subcostata* ... 75
    - *Cuphea ignea* ... 76

- **Rubiaceae** ... 77
    - *Ixora* sp. ... 77
    - *Ixora casei.* ... 79
    - *Mussaenda erythrophylla* ... 80
    - *Mussaenda philippica* ... 81
    - *Mussaenda philippica* ... 82
    - *Pentas lanceolata* ... 83

- **Amaryllidaceae** ... 84
    - *Hippeastrum rutilum* ... 84
    - *Hymenocallis littoralis* ... 85
    - *Crinum asiaticum* ... 86
    - *Zephyranthes grandiflora* ... 87

- **Turneraceae** ... 88
    - *Turnera ulmifolia* ... 88

- **Combretaceae** ... 90
    - *Combretum indicum* ... 90

- **Araceae** ... 92
    - *Anthurium andraeanum* ... 92
    - *Caladium bicolor* ... 93
    - *Epipremnum aureum* ... 95

- **Sterculiaceae** 96
  Melochia villosisima 96

- **Araliaceae** 97
  Polyscias balfouriana 97

- **Amaranthaceae** 99
  Celosia cristata 99

- **Scrophulariaceae** 100
  Angelonia salicariifolia 100

- **Melastomataceae** 101
  Melastoma candidum 101
  Melastoma dodecandrum 102

- **Lecythidaceae** 103
  Barringtonia racemosa 103

- **Iridaceae** 104
  Trimezia martinicensis 104

- **Nepenthaceae** 105
  Nepenthes sp. 105

- **Marantaceae** 106
  Ctenanthe setosa 106

- **Bignoniaceae** 107
  Pyrostegia venusta 107
  Spathodea campanulata 108

Pteridophyta

- **Huperziaceae** 111
  Phlegmariurus phlegmaria 111

- **Lycopodiaceae** 112
  Lycopodium japonicum 112

- **Aspidiaceae** 113
  Tectaria sp. 113

- **Nephrolepidaceae** 114
  Nephrolepis cordifolia 114
  Nephrolepis hirsutula 115
  Nephrolepis acutifolia 116
  Nephrolepis biserrata 117

- **Aspleniaceae** 118
  Asplenium nidus 118

- **Davalliaceae** 119
  Davallia mariesii 119
  Davallia divaricata 120

- **Vittariaceae** 121
  Vittaria ophiopogonoides 121
  Vittaria taeniophylla 122

- **Polypodiaceae** 123
  Microsorium punctatum 123

Angiospermae

● Labiatae

Plectranthus scutellarioides

English name: Coleus

Taxonomic position: Genus Plectranthus, family Labiatae

Description: Herbaceous plant. Stems are usually purple, quadrangular, puberulent and branched. Foliage is membranous, variable in colors, shape and size; leaf blades oval, yellow, dark red, purple or green, puberulent on both surfaces; apex obtuse to shortly acuminate; base broad cuneate to rounded; margin serrated or scalloped. Inflorescence is verticillaster, many flowered; corolla pale purple or blue. Nutlets are brown, often oval or oblate with lustre.

Use: Ornamental plant in the parterre or on the roadside. Also used as cut foliage for cut-flower arrangement.

Euphorbiaceae

Euphorbia milii

English name: Christ plant

Taxonomic position: Genus Euphorbia, family Euphorbiaceae

Description: Sprawling shrub. Stems are mostly branched, longitudinally ribbed, densely covered with stiff sharp conical spines. Leaves are obovate or oblong, alternate, mainly on new shoot, 1.5–5 cm long, 0.8–1.8 cm wide, sessile or subsessile; apex rounded, mucronate; base attenuate; margin entire. Stipules are subulate, very minute, caducous. Cyathia is axillary at the upper part of branches. Bracts are red, pink, yellow or white, showy. Capsules are 3-ribbed ovate, gray brown, minutely tuberculate. It blossoms and fruits throughout the year.

Use: Potted plant in public places like hotels or shops for ornamentation. Used as a thorn hedge and also as medicine.

Jatropha integerrima

English name: Peregrina, spicy Jatropha, fire cracker

Taxonomic position: Genus Jactropha, family Euphorhiaceae

Description: Shrubs or small tree with a height of 1–2 m. Leaves are alternate; leaf blades oblanceolate, usually clustered on the top of branches, acuminate at apex, adaxially dark green and smooth, abaxially purplish green. Petioles are tomentulose. Plants are monoecious. Inflorescence is cyme; flowers are unisexual, borne on different inflorescences; petals 5; corolla red. Pink variety is also available. The bloom period starts from spring to autumn.

Use: Ornamental potted flower in the courtyard or garden.

Euphorbia cyathophora

English name: Dwarf poinsettia, fire-on-the mountain, painted leaf

Taxonomic position: Genus Euphorbia, family Euphorbiaceae

Description: Perennial herb, 1 m tall. Roots are cylindrical, sometimes lignified at base. Stems are erect, usually glabrous, branched at the upper part. Leaves are alternate; leaf blades oval, elliptic or oval-elliptic, glabrous; apex acuminate or rounded; margin undulately lobed, toothed, or entire. Cyme is terminal; cyatha solitary. Involucre is campanulate, usually green. Male flowers are numerous, usually projecting beyond involucre. Appical internodes of branches are shortened at flowering; bracts are red clustered, open to four directions, similar to leaf blades, the main part for ornamentation.

Use: Potted plant or cut-flower material. Often used for background decoration in the flower border or clearing.

Acalypha hispida

English name: Chenille plant, red hot cat's tail

Taxonomic position: Genus Acalypha L., family Euphorbiaceae

Description: Evergreen bush, usually 0.5–3 m tall, upto 3 m in its origin and less than 1 m tall as a potted plant. Twigs are grayish tomentulose when young; branchlets glabrous. Foliage is papery, alternate; leaf blades broadly oval or oval, acuminate or acute at apex. Spikes are axillary, 30–60 cm long, terete, pendulous. Flowers are unisexual, pale red or dark red, small, apetalous. It blooms from February to November.

Use: Wildly used as an ornamental plant in courtyards or gardens. Common potted plant, not taller than 1 m in the greenhouse.

Codiaeum variegatum

English name: Variegated croton

Taxonomic position: Genus Codiaeum A.Juss, family Euphorbiaceae

Description: Shrubs or small trees, 2 m tall. Branches are glabrous; leaves thin, leathery, variable in size; leaf blades ovate, linear or hastate, glabrous on both surfaces, cuneate at base; leaf color ranged from green, pale green, purplish red, or in mixture of purplish-red and yellow. Green leaf blade is dotted with yellow or golden spots or stripes of spots. Petiole is 0.2–0.25 cm long. Raceme is axillary; male flower white; pedicels slender; female flower yellowish, apetalous; disk annular; pedicels slightly thick. Capules are subglobose, glabrous; seeds about 6 mm long. Bloom period starts from September to October.

Use: Common foliage plant in the garden or park in the tropical or subtropical areas.

Acalypha wikesiana

English name: Copperleaf

Taxonomic position: Genus Acalypha L., family Euphorbiaceae

Description: Shrub with a height of 1–4 m. Twigs are covered with short fine hairs. Leaves are papyraceous; leaf blades usually broad ovate, bronze-green or pale red, usually with red or purple irregular spots and patches, acuminate at apex, obtuse at base, crenate at margin, sparsely puberulous along leaf veins at lower part. Stipules are narrowly deltoid, covered with short hairs.

Use: Common foliage plants in gardens, parks or on roadsides in the tropical or subtropical areas. Used to treat some diseases like enteritis, dysentery, hematochezia and dermatitis eczema.

Leguminosae

Caesalpinia pulcherrima

English name: Peacock flower

Local name: Yapese: sowur

Kosrae: alder

Pohnpei: kiepw

Taxonomic position: Genus Caesalpinia, family Leguminosae

Description: Large shrub or small tree, up to 3 m tall. Branches are green or pinkish green, sparsely thorned. Leaves are bipinnately compound, 4 to 8 pairs, oppositely arranged; leaflets 7 to 11 pairs, oblong or obovate, oblique at base, emarginate at apex. Petiolules are quite short. Raceme is terminal or axillary; petals rounded, stipitate, pale yellow, mostly in a mixture of golden and red; pedicels 7 cm long. Pods are black; seeds 6 to 9. It flowers and fruits almost all year around.

Use: Ornamental plant in tropical areas. Seeds can be used as medicine to promote blood circulation.

Calliandra surinamensis

English name: Pink powder puff

Taxonomic position: Genus Calliandra, family Leguminosae

Description: Deciduous Shrub or small tree. Branches are expanded; twigs rough, brown. Stipule is ovate-lanceolate, persistent. Capitulum is axillary, 25–40 flowered. Peduncles are 1–2.3 cm long. Calyx is green; corolla purplish red, glabrous at the apex of lobe. Stamen is white, protruding from corolla, white; with dark red filaments at upper part. Pods are dark-brown, linear-oblanceolate, dehiscent from top to bottom; carpel revolute when ripe. Seeds are oblong and brown. Flowering period starts from August to September and fruiting period from October to November.

Use: Ornamental potted plants in the garden.

Delonix regia

English name: Royal poinciana, Flamboyant

Taxonomic position: Genus Delonix, Family Leguminosae

Description: Deciduous tree with a height of upto 20 m. Crown is broad. Leaves are bipinnate compound; leaflets oblong. Compound raceme is terminal or axillary. Flowers are big, bright red or orange red, 7–10 cm in diameter. Pods are ligneous, upto 50 cm long. It blooms in summer. National tree of The Republic of Madagascar. Municipal tree of Xiamen, Panzhihua city as well as Nanshi in Taiwan.

Use: Often planted in courtyards or on roadsides.

Balsaminaceae

Impatiens balsamina

English name: Garden balsam, Touch-me-not

Taxonomic position: Genus Impatiens, family Balsaminaceae

Description: Annual herb. Stems are erect, fleshy, less branched. Leaves are alternate, sometimes opposite at lowest part; leaf blades lanceolate or narrowly elliptic, acuminate at apex, cuneate at base, deeply serrate, with pairs of sessile black glands toward base, glabrous or puberulent on both surfaces. Flowers are axillary, solitary or in a fascicle of 2–3 flowers, pink, yellow, red, purple, white or golden. Petal is single or double. Capsules are broadly fusiform. It flowers from July to October.

Use: Ornamental plant in the parterre and courtyards or on roadsides. Used to treat rheumatism and arthralgia, and petal juice for coloring.

Apocynaceae

Plumeria rubra L. cv. *Acutifolia*

English name: Plumeria
Local name: Yapese: sawur
Kosrae: sruhsrah
Pohnpei: sawhn
Taxonomic position: Plumeria, Apocynaceae
Description: Small tree, deciduous, about 5 m tall, 15–20 cm in diameter at breast height. Branches are fleshy with milky sap. Leaf blades are thick papyraceous, oblong-oblanceolate or oblong, narrowly cuneateat base, acuminate at apex, dark green adaxially, pale green abaxially, glabrous on both surfaces.

Midveins are concave adaxially, slightly convex abaxially; lateral veins flattened on both surfaces, 30–40 on each side of the midvein, not confluent to margin to form marginal veins. Petiole is 4–7.5 cm long, glandular at upper base, glabrous. Flowers are often white, pink or bicolored with shades of yellow in the centre of flower. Red-flowered variety is a hybrid. The tree is known as "Temple Tree" or "Pagoda Tree".

Use: Ornamental potted plants in the courtyard. Also used to treat cough, fever, diarrhea, lithangiuria or to prevent sunstroke.

Rubra forma acutifolia cv. *'Gold'*

English name: Plumeria
Local name: Yapese: sawur
Kosrae: sruhsrah orangrang
Pohnpei: sawhn
Taxonomic position: Genus Plumeria, family Apocynaceae
Description: Small tree, deciduous. Leaf blades are ovate, acute at apex. Petals are red when young, yellow in bloom. Other features are the same as those of *Plumeria rubra L.* cv. *Acutifolia.*

Use: Ornamental plants in courtyards. Also used to treat cough, fever, diarrhea, lithangiuria or to prevent sunstroke.

Plumeria rubra cv. *Acutifolia*

English name: Plumeria

Local name: Yapese: sawur

Kosrae: sruhsrah

Pohnpei: sawhn

Family: Plumeria, Apocynaceae

Description: Small tree, deciduous. Leaf blades are ovate, rounded or cuspidate at apex. Petals are pure-white, yellow only at base. Other features are almost the same as those of Plumeria rubra *L.* cv. *Acutifolia*.

Use: Ornamental plants in the courtyard. Also used to treat cough, fever, diarrhea, lithangiuria or to prevent sunstroke.

Plumeria pudica

Taxonomic position: Genus Plumeria, family Apocynaceae

Description: Small tree, deciduous. Branches are robust, fleshy, green, glabrous, with milky sap. Leaves are thick papyraceous, terminally clustered; leaf blades hastate or spatulate, dark green adaxially, light green abaxially, glabrous on both surfaces; midveins, concave adaxially, slightly convex abaxially; lateral veins, flattened. Cymes are terminal; peduncles trifurcate, green, fleshy; pedicels pale red; corolla white in the outer side, tinged with pale red spots in the outer side of corolla tube and the left side of the outer side of corolla segments, yellow in the inner side. It flowers from May to October.

Use: Ornamental plant in the courtyard or garden. Used to make a hedge.

Nerium indicum

English name: Oleander

Taxonomic position: Genus Nerium, family Apocynaceae

Description: Evergreen large shrub, 5 m tall. Branches are grayish green; twigs ribbed, puberulent, caducous when old. Leaves are 3–4, in whorls; leaf blades dark green adaxially, light green abaxially; leaf midveins concave; petiole flattened. Cymes are terminal; corolla dark red or pink, funnel-shaped when single petaled and 5 lobed. Seeds are oblong. It blooms nearly throughout the year, especially in summer. The fruiting period starts from winter to spring.

Use: Ornamental plant in gardens, courtyards or on roadsides, but poisonous. The oil extracted from seeds can be used as lubricating oil.

Thevetia peruviana

English name: Yellow oleander

Taxonomic position: Genus Thevetia Linn., family Apocynaceae

Description: Evergreen tree plant, glabrous, upto 5 m tall. Barks are brown; leaves leathery, alternately arranged; leaf blade linear or linear-lanceolate. Cymes are terminal; flower yellow, big, fragrant. Drupe is compressed triangular-globose. It blooms from May to December.

Use: Landscape plant in gardens or courtyards, but poisonous.

Catharanthus roseus

English name: Rosy periwinkle

Taxonomic position: Genus Catharanthus, family Apocynaceae

Description: Subshrub, upto 60 cm tall, glabrous or only puberulent. Stems slightly branched, containing fluid, square-shaped, striate, grayish green; internodes 1–3.5 cm long. Leaf blades, membranous, obovate-oblong; apex minutely apiculate; base broadly cuneate to cuneate, attenuate to form petiole. Leaf veins are oblate adaxially, slightly convex abaxially; lateral veins about 8 pairs. Cymes are axillary or terminal, 2–3 flowered; calyx 5-partite; corolla red, salverform; corolla tube cylindrical, constricted at throat, setose. Stamens are inserted on upper corolla tube; anther included in the throat, separated from stiyma. Flowers are glossy, white, pink, red, yellow or in other colours.

Use: Landscape or ornamental plant in courtyards. It can be used to kill pain, relieve sleeplessness and inflammation, or relax bowels.

Allemanda neriifolia

English name: Bush allamanda

Taxonomic position: Genus Allamanda, family Apocynaceae

Description: Evergreen shrub with climbing or erect stems. Leaves are 3–5 whorled; leaf blades elliptic or oblanceolate oblong, pubescent. Inflorescence is cyme; petals goldenyellow, with orange red stripes on the throat. Corolla is funnel-shaped, 5 lobed, overlapping left or right; base ampliate, with 5 stamens inside. Capsules are spherical with long prickles. It flowers from May to June.

Use: Landscape plant in courtyards or gardens. Ornamental potted plant in living rooms, balconies or parks.

Allamanda violacea

English name: Violet Allamanda

Taxonomic position: Genus Allamanda, family Apocynaceae

Description: Evergreen shrub with climbing or erect stems containing milky sap. Leaves 4 whorled; leaf blades oblong or obovate-lanceolate. Inflorescence is axillary; flowers funnel-shaped, dark pink or purplish red. Corolla is 5-lobed. It blooms from late spring to autumn.

Use: Landscape planted in courtyards or gardens. Ornamental potted plant in living room, balconies or parks.

Cryptostegia sp.

English name: Rubber vine

Taxonomic position: Genus Cryptostegia, family Apocynaceae

Description: Woody perennial vine. It contains latex and is thus named rubber vine.

Use: Landscape or ornamental plant in courtyards, gardens or on roadsides. It is a potential rubber producing plant.

Zingiberaceae

Curcuma longa

English name: Turmeric

Taxonomic position: Genus Curcuma, family Zingiberaceae

Description: Perennial herb with a sturdy rhizome. Roots are thick, tuberous at tip. Leaf blades are oblong to elliptical, short acuminate at apex. Scape appears from leaf sheath. Spike is terete; bracts pale green, oval or oblong; upper bracts white, flowerless inside, tinged with light red on margin. Corolla is pale yellow or white. It flowers in August.

Use: Ornamental plant in gardens or courtyards. The rhizome serves as a Chinese traditional medicinal herb and is also used as a food colorant.

Curcuma australasica

Taxonomic position: Genus Curcuma, family Zingiberaceae

Description: Perennial herb. Bracts are green or pale green, purplish red at apex, white at base; upper bracts purplish red or streaked with white and pink. Other features are almost the same as those of *Curcuma longa*.

Use: Ornamental plant in courtyards or gardens.

Zingiber zerumbet

English name: Bitter ginger, Shampoo ginger

Taxonomic position: Genus Zingiber Adans., family Zingiberaceae

Description: Perennial herb. Rhizome is tuberous. Plants can grow up to 2 m tall. Leaf blades arelanceolate or oblong-lanceolate, sessile or shortly petiolate. Inflorescence is usually conical, obtuse at apex. Bracts are densely imbricate, subrounded, pale green when young, red, membranous at margin. Calyx is membranous; corolla tube slender; corolla lobe lanceolate.Labellum is pale yellow; lateral lobe obovate. Capsule is oval; seeds black. It blooms from July to September.

Use: Cut-flower or ornamental flower in courtyards.

Alpinia purpurata

English name: Red ginger

Taxonomic position: Genus Alpinia, family Zingiberaceae

Description: Herb. Plants are robust, 0.8–2.5 m tall. Leaf blades are medium-sized, lanceolate, glabrous, caudately mucronate at apex. Panicle is 15–30 cm long. Bracts are peltate; bracteoles shell-like; both bracts and bracteoles red or purplish red.

Use: Ornamental flower in parks or courtyards.

Hedychium coronarium

English name: White ginger lily

Taxonomic position: Genus Hedychium Koenig, family Zingiberaceae

Description: Herb, 1–2 m tall. Leaves are alternate; leafblades long and narrow, mucronate at both ends, glabrous adaxially, sparsely puberulous abaxially. Inflorescence is spike; calyx tubulose. Bracts are imbricate, ovate, green. Petals are white, yellow at base, bifid at apex. It grows in a humid or warm environment.

Use: Ornamental plant in courtyards or on wetland. Rhizome and fruit are used as medicine.

Hedychium coronarium

English name: Orange ginger lily, Scarlet ginger lily

Taxonomic position: Genus Hedychium Koenig, family Zingiberaceae

Description: A red flower variety. Herb, 1–2 m tall. Leaves are alternate; leaf blades, long and narrow, mucronate at both ends, glabrous adaxially, sparsely puberulous abaxially. Inflorescence is spike; calyx tubulose; bracts imbricate, oval, green; petal pale red.

It grows in a humid or warm environment.

Use: Ornamental plant in courtyards or on the wetland. Rhizome and fruit are used as medicine.

Costus spciousus (J.Koenig) Sm.

English name: Canereed

Taxonomic position: Genus Costus, family Zingiberaceae

Description: erennial herb, 1–3 m tall. Stems are ligneous at base, usually branched at apex, spirally twisted. Leaf blades are oblong or lanceolate, acuminate at apex, densely sericeous abaxially. Spike is terminal. Bracts are oval, leathery, red; bracteoles 1.2–1.5 cm long, pale red. Calyx is leathery, red; corolla white, or red at apex; labellum broad trumpet-shaped, pure-white. Seeds are black, glossy. It flowers from July to September and fruits from September to November.

Use: Ornamental plant under trees in tropical gardens. Can be used as cut-flower.

Malpighiaceae

Tristellateia australasiae

English name: Shower of Gold, Maiden's jealousy

Taxonomic position: Genus Tristellateia, family Malpighiaceae

Description: Evergreen woody vine, 10 m long. Leaves are opposite, papyraceous or subleathery; leaf blades ovate, acute to acuminate at apex, rounded to cordate at base, entire. Raceme is terminal or axillary; flowers yellow. Samaras are stellate. It flowers in August and fruits in October.

Use: Ornamental plant. Used to treat anorexia, indigestion, bruise or pyogenic infections.

Malvaceae

Hibiscus rosa - sinensis

English name: China rose, Chinese hibiscus

Local name: Yapese: Folores

Pohnpei: kolou

Taxonomic position: Genus Hibiscus, family Malvaceae

Description: Evergreen shrub, 1–3 m tall. Branchlets are terete, sparsely stellate pilose. Leaf blades are ovate, pilose only near the abaxial veins. Flowers are solitary, axillary on upper branches, usually pendulous. Corolla is funnel-shaped, 6–12 cm in diameter, white, red, yellow or pink. Capsule is oval, 2.5 cm long, smooth, glabrous, beaked. Flowers are double petaled in some varieties. It blooms all the year around.

Use: Ornamental plant in gardens, especially in tropical or subtropical areas.

Hibiscus mutabilis

English name: Cotton rose

Taxonomic position: Genus Hibiscus, family Malvaceae

Description: Deciduous shrub or small tree. Leaf blades are broadly ovate to ovate or cordate. Flowers are solitary, mostly double-petaled, axillary on upper branches, dark red but initially white or light red. Capsule is oblate, yellowish hispid and sericeous. Seeds are reniform, villous abaxially.

Use: Ornamental plant in courtyards or parks.

Campanulaceae

Isotoma axillaris

Taxonomic position: Genus Isotoma, family Campanulaceae

Description: Perennial erect herb, 50–80 cm tall, containing latex. Leaves are alternate, papery; leaf blades lanceolate. Flowers are solitary, axillary; corolla tube long, white, 5-lobed; capsule elliptic.

Use: Ornamental potted plant on balconies or roadsides, or in parterre.

Asteraceae

Zinnia elegans

Taxonomic position: Genus Zinnia, family Asteraceae

Description: Annual herb. Stem is upright with rough hairs. Leaf blades are broadly ovate or oblong, slightly amplexicaul at base, coarse on both surfaces. Capitulum is solitary on top of branches. Bracts are multi-layered, with black margin. Ligulate flowers are dark red, rose or white; tubulose flowers yellow or orange; lobes at apex, ovate-lanceolate, densely tawny tomentose. It blooms from June to September. Single petal, double petal, curled-leaf, rugose-leaf or different-colored flowers are found in garden cultivars.

Use: Ornamental plant in parterre or gardens, or house plant.

● Taccaceae

Tacca leontopetaloides

English name: East indies arrowroot, Pia

Taxonomic position: Genus Tacca, family Taccaceae

Description: Perennial herb. Leaf blade is obovate or ovate, 3 palmately lobed; lobes pinnatipartite. Inflorescence is umbel; flowers pale yellow, yellowish green or purplish green.

Use: Rare ornamental plant. Tuberous roots are poisonous, inedible.

Acanthaceae

Barleria cristata

English name: Philippine violet, Bluebell barleria

Taxonomic position: Genus Barleria, family Acanthaceae

Description: Small shrub. Stems are terete, pubescent, branched. Leaves are papery; leaf blades ellipsoid, oblong ellipsoid or ovate. Short branches are branched. Flowers are densely clustered on short branches; Bracts are foliaceous, sessile. Bracteoles linear or lanceolate; apex acuminate, mucronate, sometimes serrate; teeth spiny at apex. Corolla is bluish purple or white; corolla tube cylindric, gradually enlarged at throat; lobes subequal, oblong. Ovary is oblate, oblong elliptic, glabrous. Disk is cup-shaped, included in the lower part of ovary. Style is linear, glabrous; stigma slightly ampliate. Capsule is oblong, glabrous, acute at both ends. It blooms from November to December.

Use: Ornamental plant in the park. It can be used as medicine.

Pseuderanthemum reticulatum

English name: Yellow-vein Eranthemum, Golden Pseuderanthemum

Taxonomic position: Genus Pseuderanthemum, family Acanthaceae

Description: Perennial herb, 0.5–2 m tall. Leaves are oppositely arranged; leaf blades broadly lanceolate to oblanceolate, with irregular incised margin, golden when newly emerge, yellowish green or emerald. Golden speckles are scattered around leaf margin and named as such. Flowers are terminal, red or white. It blooms from spring to summer.

Use: Longitudinally planting or clustered planting in courtyards. Indoor potted plant.

Pseuderanthemum reticulatum

Taxonomic position: Genus Pseuderanthemum, family Acanthaceae

Description: Perennial herb, 0.5–2 m tall. Leaves are opposite, broadly lanceolate to oblanceolate, with irregular incised margin, red along veins of new leaves, green or dark green. Flowers are terminal, red or white. It blooms from spring to summer.

Use: Indoor potted plant. Longitudinally planting or clustered planting in courtyards.

Thunbergia grandiflora

English name: Bengal trumpet, Blue skyflower

Taxonomic position: Genus Thunbergia, family Acanthaceae

Description: Climbing herb or shrub, rarely erect, glabrous or puberulent. Leaves are opposite, petiolate; leaf blades ovate, lanceolate, cordate or hastate, acute or acuminate at apex, sometimes rounded at apex, pinnately, palmately or ternately veined. Flowers are solitary or in racemes, terminal or axillary, usually big and showy; corolla funnel-shaped; corolla tube short, curved inward or obliquely; throat widened; limb spreading. Pollen grains are spherical; disk annular, or cushion-shaped.

Use: Ornamental plant in courtyards or parks.

Staurogyne concinnula

Taxonomic position: Genus Staurogyne, family Acanthaceae

Description: Herb. Stem is very short, densely pilose. Leaves are opposite, fascicled in a basal rosette; Petiole is quite short, pubescent. Leaf blade is spatulate-oblong or spatulate-lanceolate, obtuse to rounded at apex, attenuate at base, abaxially glaucous and sparsely puberulous, pilose along leaf veins. Raceme is terminal or subterminally axillary, sparsely flowered. Peduncle and rachis are slender, puberulent. Bract is spatulate-linear; bracteole linear, subequal to bract. Calyx is yellow or white; corolla red, aromatic.

Use: Ornamental plant.

Gesneriaceae

Alloplectus martius

Taxonomic position: Genus Alloplectus, family Gesneriaceae

Description: Evergreen perennial herb. Flowering bulb. Leaves are oppositely arranged, broadly lanceolate, adaxially brownish green and lustrous, abaxially purplish brown; margin serrated; veins conspicuous, finely crinkled. Flowers are terminal or axillary; florets tubular, orange. Stem is erect, quadrangular, yellowish green, slightly transparent. Umbel is axillary, pentagonally cupular; apex acute, 3 toothed, carmine. Flowers are tubular, 5 petaled, semi-rounded, golden red.

Use: Ornamental plant in gardens, parterre or courtyards in temperate areas. The seed oil serves as medicine. Raw material of alkyd resin.

Episcia cupreata

English name: Flame violet

Taxonomic position: Genus Episcia, family Gesneriaceae

Description: Evergreen perennial herb, prostrate and many branched. Leaves are oppositely arranged; leaf blades elliptical, dark green or brown with serrated margin and cordate base; adaxial surface rugose, densely tomentulose, white on midvein from base to apex, grayish green along midvein and lateral veins; abaxial surface pale green or pale red. Stolon arises from basal axil, shooting from soil to form a plantlet at apex. Flowers are solitary or axillary in a small fascicle, usually bright red. It blooms from spring to autumn.

Use: Indoor ornamental plant.

● Orchidaceae

Taeniophyllum sp.

Taxonomic position: Genus Taeniophyllum, family Orchidaceae

Description: Small leafless plant. Roots are well developed, numerous, clustered, slightly flattened, recurved, 2–10 cm long, 0.6–1.2 mm in diameter, borne on tree bark, spreading like spiders.

Use: Indoor potted plant or plant on drift wood.

Rodriguezia sp.

Taxonomic position: Genus Robiquetia Gaud, family Orchidaceae

Description: Stems are sturdy, terete; internode about 2 cm; roots well developed, branched on lower nodes. Leaves are distichous; inflorescence opposite to leaf. Panicles are densely many small flowered; flowers yellowish green. It flowers from June to September.

Use: Indoor potted plant or plant on drift wood.

Epidendrum radicans

Taxonomic position: Genus Epidendrum, family Orchidaceae

Description: Stem is slender, upto 1 m long. Raceme is terminal; flowers orange; labellum yellow at base, dotted with dark red spots, 4 lobed, with lacerate lateral lobes.

Use: Potted plant, or landscape plant in courtyards.

Aranda Hybrid

Taxonomic position: Hybrid of Genus Vanda, family Orchidaceae

Description: Epiphytic orchid in tropical areas. Leaves are alternately arranged, thickly leathery. Inflorescence is relatively long, axillary; flowers red; labellar base and gynandrium yellow.

Use: Ornamental potted flower. Or used as cut-flower.

Arachnis Maggie Oei

Taxonomic position: Genus Vanda, family Orchidaceae

Description: Epiphytic monopodial orchid, with conspicuous stem and aerial roots. Aerial roots are well-developed, fleshy; leaf blades semiterete, arranged in two rows on both sides of stem.

Use: Potted flower. Or used as cut-flower.

Papilionanthe Miss Joaquim

Taxonomic position: Genus Papilionanthe, family Orchidaceae

Description: Plant is upto 2 m; stem slender; leaf blade nail-like; raceme up to 12 flowered; both 2 petals and calyx rose-purple at upper part; calyx at low part on two sides purplish. National flower of Singapore.

Use: Potted flower. Or used as cut-flower.

Spathoglottis plicata

English name: Large purple orchid

Taxonomic position: Genus Spathoglottis B, family Orchidaceae

Description: Pseudobulb is oblate-globose with leathery scaly sheath and 1–3 terminal leaves. Leaves are lorate or narrowly lanceolate, acuminate at apex, attenuate into thin petioles at base. Scape is slender or thick, densely pubescent, covered with several tubulose sheaths subtending peduncle at lower part. Inflorescence is raceme; sepal lanceolate or ovate-lanceolate, pubescent; pedicel and ovary densely pubescent; petalwide, oblong, equal to sepal; labellum subequal to petal, rounded or truncate at apex; flower purplish red.

Use: Ornamental potted flower. Pseudobulb is used as medicine to treat cough, bruise, expectoration or phthisis.

Spathoglottis micronesia

Taxonomic position: Genus Spathoglottis B, family Orchidaceae

Description: Pseudobulb is oblate-globose with scaly leathery sheath and 1–3 terminal leaves. Leaves are linear or narrowly lanceolate, acuminate at apex, attenuate to slender petioles at base. Inflorescence is raceme; flowers white; sepal, big and broad; petal broad rounded; labellum slightly short, truncate at apex, yellow at base; scape slender or robust, densely pubescent, surrounded by tubular sheaths subtending peduncles.

Use: Ornamental potted flower.

Dendrobium sp.

Taxonomic position: Genus Dendrobium Sw., family Orchidaceae

Description: Stems are erect or pendent, fleshy, thick, slightly oblate-terete, slightly curved at upper part, conspicuously attenuate at base, unbranched, nodular; node occasionally ampliate; internode obconical; leaves leathery, oblong, with amplexicaul sheaths at base.

Use: Ornamental flower. A few varieties can be used as medicine.

Polygonaceae

Antigonon leptopus

English name: Coral vine, Bee bush

Taxonomic position: Genus Antigonon, family Polygonaceae

Description: Evergreen woody vine. Roots are thick; stems climbs 10 m long or more; flowers mostly densely borne in a raceme as bunches, showy, slightly fragrant. It blooms from March to December.

Use: Rare ornamental flower in summer.

Agavaceae

Cordyline fruticosa

English name: Palm lily, Cabbage palm, Good luck plant
Taxonomic position: Genus Cordyline, family Agavaceae
Description: Shrubby plant, erect, 1–3 m tall. Leaves are clustered on upper part of stem or branches, oblong to oblong-lanceolate, 25–50 cm long, showy, colorful, green, golden or purplish red. Petioles are sulcate, 10–30 cm long; base broad, amplexicaul. Panicles are 30–60 cm long; bracts at the base of lateral branch big, with 3 bracts each flower. Flowers are about 1 cm long, pale red, bluish purple to yellow. Pedicels usually very short, rarely 3–4 mm long. Outer tepals at lower part closely touch inner tepals to form a perianth tube, curved outward or reflexed at upper part at anthesis. Stamen is borne in the throat of the perianth tube, slightly shorter than tepal. Style is slender, long Bloom period starts from November to next March.

Use: Indoor potted plant. Ornamental plant or landscape plant in courtyards, parks or on roadsides.

Musaceae

Heliconia spp.

English name: False bird-of-paradise

Taxonomic position: Genus Heliconia, family Strelitziaceae

Description: Evergreen perennial herb. Stems are composed of tightly rolled leaf sheaths, clustered above ground. Leaves are petiolate, banana leaf-like. Bracts red or yellow; inflorescencerrect or pendulous, mostly terminal, axillary for a few; bavicular bracts borne on both sides of inflorescence, mostly showy; sepal projecting out of bracts, like a scorpion's tail. Each flower contains 3 sepals and 3 petals; petals holding together like a tube.

Use: Ornamental plant or landscape plant. It also can be used for ikebana in living rooms or hotels.

Verbenaceae

Lantana camara

English name: Big-sage, Wild-sage, Tickberry

Taxonomic position: Genus Lantana, family Verbenaceae

Description: Evergreen shrub, 1–2 m tall. Branches are sometimes rattan-like, square, pubescent, mostly covered with short, barb-like prickles; leaves opposite, oval-oblong, acuminate at apex, rounded at base, rough with hairs on both surfaces, strongly odorous when crushed. Umbel is axillary, borne at upper part of twigs, 20 or more flowered; corolla tube slender, mostly 5-lobed at apex like a plum flower;corolla variable in color, yellow, orange pink, or dark red. Flowering stage is long. Fruit is berry, globose, purplish black when ripe.

Use: Ornamental plant in courtyards or on roadsides. Used to treat fever, tendon injury, dermatitis, eczema or diarrhoea with other herbs.

Duranta repens

English name: Golden dewdrop, Pigeon berry

Taxonomic position: Genus Duranta, family Verbenaceae

Description: Shrub, 1.5–3 m tall. Branches are usually pendulous; twigs puberulent. Leaves are opposite;petiole about 1 cm long, puberulent; leaf blades papery, oval, serrated above the middle of margin. Raceme is terminal or axillary, often conical; calyx tubulose, puberulent, 5 ribbed, 5-lobed at apex, twisted at apex when fruitset; corolla blue or bluish purple, longer than style; ovary glabrous. Drupe is globose, about 5 mm in diameter, reddish yellow and lustrous when ripe, included in persistent calyx. Flowering and fruiting period starts from May to October.

Use: Landscape plant. Used as a hedge. Used to treat diarrhea caused by malaria, to promote blood circulation or to relieve pain.

Clerodendrum thomsonae

English name: Bleeding glory-bower

Taxonomic position: Genus Clerodendrum, family Verbenaceae

Description: Shrub. Young branches are quadrangular, yellowish brown tomentose, glabrous when old. Leaf blades are papery, narrowly ovate or ovate-oblong, acuminate at apex, subrounded at base, entire. Cymes are axillary or pseudo-terminal, bifurcate; bracts narrowly lanceolate; calyx white, connate at base, ampliate in middle; calyx segments deltoid-ovate, acuminate at apex; corolla dark red, covered with glandular hairs in outer part; corolla segments elliptical. Drupe is subglobose; exocarp brownish black, glossy; persistent calyx not dilated, reddish purple.

Use: Ornamental plant in greenhouses, courtyards, parks or scenic spots. Heat-clearing and detoxifying effect. Used to relieve swelling and pain.

Clerodendrum japonicum

Taxonomic position: Genus Clerodendrum, family Verbenaceae

Description: Deciduous shrub. Branchlets are quadrangular, tomentulose; leaf blades rounded, acuminate at apex, cordate at base, sparsely short serrulate at margin. Dichasium is terminal; raceme borne on the last lateral branches of inflorescence. Bracts are broad ovate or lanceolate; calyx red, pubescent; corolla red, occasionally white. It blooms from May to November.

Use: Ornamental potted flower in courtyards or on roadsides. Used to relieve rheumatic pains and swellings.

Holmskioldia sanguinea

English name: Chinese hat plant, Mandarin's hat

Taxonomic position: Genus Holmskioldia, family Verbenaceae

Description: Evergreen shrub. Branchlets are quadrangular, 4 sulcate, puberulent. Leaves are oppositely arranged; leaf blades membranous, ovate or broad ovate, serrated at margin, with sparse hairs and glandular spots on both surfaces; petiole sulcate. Cymes are 3 flowered with 2–6 cymes forming a conical shape. Calyx is vermeil or orange red, expanded from base to form a broad obconical disk; disk about 2 cm in diameter. Corolla is red; corolla tube glandular. It blooms from late winter to early spring.

Use: Potted flower. Landscape plant in gardens.

● Cannaceae

Canna indica

English name: Indian shot, Edible canna

Taxonomic position: Genus Canna, family Cannaceae

Description: Plant is green. Leaf blade is oval-oblong. Raceme is sparsely flowered, slighlty taller than leaf blade. Flower is solitary, red; bract ovate, green; sepals 3, lanceolate, green or sometimes tinged with red; corolla tube shorter than 1 cm; corolla segment lanceolate, green or red; outer staminodes 2–3, bright red, oblanceolate in two, small in one; labellum lanceolate, curved; fertile stamen 2.5 cm; anther cell 6 mm long; styles oblate, with half of them confluent to filaments of fertile stamens.

Use: Ornamental potted plant in parterre. Used as medicine. Fiber from stems and leaves can be used to prepare staple rayon, sacks or ropes. Leaf residues left from extraction of essential oil from leaves can be used as raw material for paper making.

Oleaceae

Jasminum grandiflorum

English name: Royal jasmine, Catalan jasmine

Taxonomic position: Genus Jasminum, family Oleaceae

Description: Climbing shrub, 1–4 m tall. Branchlets are terete, or grooved. Leaves are opposite, pinnatipartite or 5–9 lobed; leaflets ovate or oblong; terminal leaflet usually narrowly rhombic. Cyme is terminal or axillary, 2–9 flowered; pedicels in the middle of cyme obviously shorter than others; corolla white, salverform. Flowers are showy, fragrant. It blooms from August to October. National flower of Pakistan.

Use: Ornamental plant.

Vitaceae

Leea indica

Taxonomic position: Genus Leea van Royen ex Linn., family Vitaceae

Description: Erect shrub. Branchlets are terete, with obtuse longitudinal ridges, rusty pubescent when tender, caducous. Leaves are simple pinnate or bipinnately compound; leaflets elongate elliptical or elliptical-lanceolate, caudate or acuminate at apex, rounded at base, irregularly serrated at margin, adaxially glabrous and green, abaxially pale green and rusty pubescent. Inflorescences are opposite to leaves, compound dichasial or umbelliform. Corolla lobe is 1.8–2.5 mm long, oval, glabrous. Flowers are bright red, and bloom from April to June.

Use: Clustered planting under trees in parks.

Lythraceae

Lagerstroemia micrantha

English name: Jakaranda

Local name: Kosrae: Mieranfha Fel

Taxonomic position: Genus Lagerstroemia, family Lythraceae

Description: Small tree or shrub. Branches are terete, glabrous; leaf blades papery, elliptical or ovate, acute or acuminate at apex, attenuate or subrounded at base, usually not equal on both sides, adaxially black brown, abaxially pale, minutely puberulent when tender, scattered pubescent along leaf midveins, densely gray puberulent. Flowers are small, dark pink or pink; floral buds globose, repand at margin. Stamens are numerous, subequal; filaments about 5 mm long. Ovary is subglobose, glabrous; style 3–5 mm long.

Use: Landscape plant in scenic spots. Potted flower and ornamental tree.

Lagerstroemia subcostata

English name: Jakaranda

Local name: Kosrae: Mieranfha Fel

Taxonomic position: Genus Lagerstroemia, family Lythraceae

Description: Deciduous tree or shrub with a height of 14 m. Bark is thin, grayish white or tea-brown, glabrous or sparsely short hirsute. Leaves are membranous; leaf blades oblong, oblong-lanceolate, rarely ovate, acuminate at apex broadly cuneate at base, usually adaxially glabrous or sometimes scattered with minute pubescence, abaxially glabrous or villous, or pubescent along midveins, sometimes in a tuft of pubescence between axil and vein. Flowers are small, white or rose; capsules elliptical; seeds winged.

Use: Landscape plant in courtyards or on roadsides. Wood is hard and highly dense in structure, and good for making furniture, blockboard, crosstie or buidling. The flower serves as medicine with the effect of detoxication and blood stasis removal.

Cuphea ignea

English name: Cigar plant, Cigar flower

Taxonomic position: Genus Cuphea, family Lythraceae

Description: Plant is 30–40 cm tall, but potted plant about 10 cm tall to flower. Leaves are papery, opposite; leaf blades emerald, lanceolate, entire. Flowers are axillary, apetalous, formed by bright red, tubular calyx; calyx tube mouth purple, white, very elegant in shape. Each calyx can last for quite a long time. It flowers almost throughout the year, especially in summer.

Use: Ornamental herb in courtyards, parterre or gardens.

Rubiaceae

Ixora sp.

English name: Ixora

Local name: Yapese: gashiyaw

Kosrae: kalsruh

Pohnpei: kefieu whorled

Taxonomic position: Genus Ixora, family Rubiaceae

Description: Shrub, 0.8–2 m tall. Branchlets are initially dark brown and lustrous, gray when old. Leaves are opposite, lanceolate, oblong-lanceolate to oblong-oblanceolate, sometimes 4 whorled due to short internode, obtuse or rounded at apex, rounded at base. Midveins are slightly concave adaxially, confluent to each other near margin; horizontal veins sporadic. Petiole is short, thick. Inflorescence is terminal, many flowered, short pedunculate; calyx tube 1.5–2 mm long; calyx limb 4 lobed with very short lobes; corolla red, pink, white, yellow or orange. 4-lobed at apex; corolla lobes obovate or subrounded; filaments quite short; anther oblong, about 2 mm long. Fruit is subglobose, twin, with a groove in the center, reddish black when ripe. It blooms from May to July. National flower of Myanmar.

Use: Landscape plant in courtyards or on roadsides. Used to make a hedge.

Ixora casei.

Taxonomic position: Genus Ixora, family Rubiaceae

Description: Small tree, 1.5–2.5 m tall. Branches are long and soft like vine. Leaves are opposite, 15–25 cm long; internodes are not reduced; petioles relatively long. Other features are almost the same as those of ixora sp. Common plant on the forest edge in Federated States of Micronesia.

Use: Ornamental plant.

Mussaenda erythrophylla

English name: Red flag bush

Taxonomic position: Genus Mussaenda, family Rubiaceae

Description: Evergreen, semi-deciduous erect or climbing shrub, 1–1.5 m tall. Leaves are papery; leaf blades lanceolate-elliptical, 7–9 cm long, 4–5 cm wide, acuminate at apex, attenuate at base, sparsely puberulent on both surfaces; veins red. Inflorescence is cyme; corolla yellow, 5 lobed, stellate; one corolla segment (sepal) enlarged to a leaflike shape, dark red, oval, 3.5–5 cm long; corolla segments acute at apex, red pubescent, with 5 longitudinal veins.

Use: Ornamental plant in courtyards, parks or on roadsides.

Mussaenda philippica

Taxonomic position: Genus Mussaenda, family Rubiaceae

Description: Evergreen semi-deciduous shrub with its flowering period from summer to autumn. Foliage is opposite; leaf blades oval-lanceolate , acuminate or acute at apex, obtuse or attenuate at base, entire at margin. Cyme is terminal; calyx 5-partite; calyx segments linear; bracts usually 1–2, big, foliaceous, rounded or broadly oval, white or yellowish white. It blooms from May to October.

Use: Ornamental plant in courtyards and parks, or landscape plant on roadsides.

Mussaenda philippica

Taxonomic position: Genus Mussaenda, family Rubiaceae

Description: Evergreen semi-deciduous shrub with its flowering period from summer to autumn. Foliage is oppositely arranged; leaf blades oval-lanceolatea, acute or acuminate at apex, obtuse or attenuate at base, entire. Cyme is terminal; calyx 5-lobed; calyx segments linear; bracts usually 1–2, big, foliaceous, rounded or broad oval, pink or pinkish white. It blooms from May to October.

Use: Ornamental plant in courtyards and parks, or landscape plant on roadsides.

Pentas lanceolata

English name: Egyptian starcluster

Taxonomic position: Pentas, Rubiaceae

Description: Shrub, 70 cm tall, pubescent. Leaf blade is ovate, elliptic or lanceolate-oblong, acute at apex, attenuate to short petiole at base. Cyme is terminal, densely flowered; flowers sessile, pink, pale red, crimson, white or other colored; style quite long; corolla purple or white; corolla mouth densely pubescent; corolla limb spreading. It blooms from summer to autumn.

Use: Ornamental plant in the parterre or for a certain setting due to its long flowering stage.

Amaryllidaceae

Hippeastrum rutilum

English name: Hippeastrum

Taxonomic position: Genus Hippeastrum, family Amaryllidaceae

Description: Bulb is subglobose, with stolons. Leaves are 6–8, emerged after flowering, bright green, ribbon-like. Scape is hollow, slightly oblate, covered with white powder. Flowers are 2–4, various in colors like pink, red, yellow or dark red; spathe-like bracts are lanceolate; pedicels slender; perianth tube green, tubular; tepal oblong, acuminate at apex, carmine, tinged with green, with small scales around the throat; stamens 6; filaments red; anther linear-oblong; stigma 3-lobed. It blooms in summer.

Use: Ornamental plant in living rooms, aisles or corridors. Used as cut-flower or plant cultivation in courtyards.

Hymenocallis littoralis

English name: Beach spider lily

Taxonomic position: Genus Hymenocallis, family Amaryllidaceae

Description: Perennial bulbous herb. Foliage is acute at apex, attenuate at base, dark green, multi-veined but sessile. Scape is oblate, broad at spathe-like base; flowers 3–8, borne at apex of scape, white, sessile; perianth tubes slender, different in length; tepals linear, often shorter than perianth tube; staminal cup campanulate or broadly funnel-shaped, serrated; style nearly equal in length to or longer than stamen. Flowers are greenish white, sweetly scented; flowering period from late summer to early autumn. Capsule is oval or annular, fleshy, dehiscent at maturity. Seeds are spongy, green.

Use: Ornamental plant in courtyards or gardens. Can be used as medicine.

Crinum asiaticum

English name: Spider Lily

Taxonomic position: Genus Crinum, family Amaryllidaceae

Description: Perennial robust herb. Bulb is long cylindrical. Leaves are 20–30, polystichous, often strap-shaped-lanceolate, 1 m long, 7–12 cm wide or wider; apex acuminate with an acute mucro; margin undulate, dark green. Scape is erect, subequal to leaf. Umbel is 10–24 flowered; spathe-like phyllary lanceolate; flower salverform and fragrant; perianth tube slender, greenish white; tepal linear, white or purple, attenuate to apex. It blooms in summer. Widely planted in some Buddhist countries.

Use: Ornamental plant in gardens, scenic spots, schools or offices. Often planted on the lawns of residential districts. Used as a hedge.

Zephyranthes grandiflora

English name: Pink rain lily

Taxonomic position: Genus Zephyranthes, family Amaryllidaceae

Description: Perennial herb. Bulbs are ovoid. Basal leaves are usually several clustered, linear, flattened. Flowers are solitary, borne at apex of scape, with spathe-like involucres at lower part. Involucral bracts are usually purplish red, connate at lower part to form tubes. Flowers are rose or pink; tepals 6 lobed; tepal lobes obovate, slightly pointed at apex. It blooms from spring to autumn.

Use: Ornamental plant in gardens. It can be used as border material in parterre or footpaths. It also serves as medicine for detoxification and poor blood circulation.

Turneraceae

Turnera ulmifolia

English name: Yellow alder

Local name: Yapese: sowur

Kosrae: alder

Pohnpei: kiepw

Taxonomic position: Genus Turnera, family Turneraceae

Description: Evergreen liana, originated from the tropical jungle of South America. Leaves are ovate, alternate, acuminate at apex, serrated at margin; whorled due to short internode. Corolla is yellow or white; sepals, petals and filaments all 5 each; anther strip-shaped. It often blooms in the morning and closes petals in the evening.

Use: Ornamental plant in parks or on roadsides.

Combretaceae

Combretum indicum

English name: Chinese honeysuckle, Rangoon creeper
Taxonomic position: Genus Combretum, family Combretaceae
Description: Climbing shrub, 2–8 m tall. Branchlets are brownish yellow pubescent. Leaves are opposite; leaf blades membranous, ovate or elliptical, short acuminate at apex, obtuse at base, glabrous adaxially, sometimes sparsely brown puberulent abaxially, densely ferruginous puberulent when young. Spikes are terminal, clustered in corymb. Bracts are oval to linear-lanceolate, puberulent. Fruits are distinctly sharply 5-ridged; exocarp crispy, thin, bluish black or chestnut-colored when ripe. Seeds are white, cylindrical-fusiform. It flowers in early summer and fruits in late autumn.

Use: Ornamental plant in courtyards. Used as a hedge. Seeds are the best ascaricide in traditional Chinese medicine.

● Araceae

Anthurium andraeanum

English name: Tailflower, Painter's palette, Flamingo flower
Taxonomic position: Genus Anthurium, family Araceae
Description: Evergreen perennial herb. Stem internode is short. Foliage is basal, green, leathery; leaf blades oblong-cordate or oval-cordate, entire. Petioles are slender; spathe leathery, oval-cordate, waxy, orange red or scarlet; spadix yellow, 5–7 cm long. It blooms frequently in a year. Flowers are long lasting.

Use: Used as cut-flower. It can absorb exhausted or poisonous gas like ammonia and formaldehyde. Meanwhile, it serves as an air humidifier.

Caladium bicolor

English name: Elephant ear, Angel wings

Taxonomic position: Genus Caladium Vent., family Araceae

Description: Evergreen herb. Tuber is depressed-globose. Petioles are glossy, covered with white powder at upper part. Leaf blades are hastate-ovate to oval-triangular, adaxially dotted with transparent or non-transparent spots, abaxially pinkish green. Peduncles are shorter than petioles. Spathe tubes are ovate, green at outer side, greenish white at inner side, bluish purple at base; limbs protuberant, white; spadix attenuate towards both ends.

Use: Ground cover plant, shade-tolerant, often growing in some temperate areas. Ornamental potted plant in gardens. Tubers can be used as medicine.

Epipremnum aureum

English name: *G*olden pothos, Hunter's robe, Money plant

Taxonomic position: Genus Epipremnum, family Araceae

Description: Tall liana. Stems are climbing, many branched; internodes longitudinally sulcate. Branches are pendulous; twigs whip-shaped, slender; leaf sheath long; leaf blades thin, leathery, emerald, ovate or ovate-oblong, usually dotted with irregular pure-yellow speckles (especially on upper surface), short acuminate at apex, cordate at base, entire.

Use: Foliage plant in front of hotels or in lobbies. Indoor flower. It can absorb impurities and purify air.

Sterculiaceae

Melochia villosisima

English name: Juteleaf melochia

Taxonomic position: Genus Melochia, family Sterculiaceae

Description: Semi-shrub. Branches are yellowish brown, slightly covered with stellate pubescence; leaves thin, papery, ovate or lanceolate, with serrated margin, nearly glabrous adaxially, pubescent abaxially. Cyme is terminal or axillary; bracteoles linear, within cyme; petals 5 each flower, oblong, initially white and then pale red. It blooms from summer to autumn.

Use: Ornamental plant in courtyards.

Araliaceae

Polyscias balfouriana

English name: The geranium aralia

Taxonomic position: Genus Polyscias, family Araliaceae

Description: Plants are less branched. Leaves are odd-pinnately compound, alternate; leaflets green, various in shape and number, ovate to lanceolate, with serrated or divided margin, short petiolate. Umble is conical; flowers small, numerous, green. Twigs are soft; leaves 1-pinnate; leaflets 2–4 pairs, green, glossy, edged in white.

Use: Indoor foliage plant; high adaptability to indoor environment.

Amaranthaceae

Celosia cristata

English name: Cockscomb

Taxonomic position: Genus Celosia, family Amaranthaceae

Description: Annual herb, glabrous. Stem is erect with distinct strips, branched; leafblade ovate or lanceolate; spike fleshy, flattened, often cristate or convolute; flowers clustered, numerous; perianth, usually red, purple, yellow, orange or in red and yellow. It blooms from July to September.

Use: Ornamental plant in gardens, courtyards or parterre. Both flower and seed can be used as astringent.

Scrophulariaceae

Angelonia salicariifolia

Taxonomic position: Genus Angelonia, family Scrophulariaceae

Description: Perennial herbaceous flower, 30–70 cm tall. Petals are labiate, 4-lobed at upper part; flower purple, light purple, pinkish purple or white; capsules and seeds quite small. It blooms all the year round, especially in spring, summer and autumn.

Use: Ornamental plant in parterre and on flower-stand. It can be used for aquatic plant cultivation or tea drinking.

Melastomataceae

Melastoma candidum

Taxonomic position: Genus Melastoma, family Melastomataceae

Description: Evergreen shrub. Leaves are opposite, broad ovate, puberulent on both surfaces, actue at apex, cordate at base, entire. Inflorescence is corymb; flowers bisexual, white, clustered on the top of branches. It blooms from May to July.

Use: Wild plant. Foliage plant on roadsides or in parks.

Melastoma dodecandrum

English name: Twelve stamen melastoma herb

Taxonomic position: Genus Melastoma, family Melastomataceae

Description: Pendulous or procumbent subshrub. Stems are branched; leaves opposite; leafblades ovate or elliptic, strigose on adaxial margins and abaxial veins; main veins 3–5. Cyme is terminal, with leaf-like involucre at base; flowers bisetual, pale purple. It blooms from May to July.

Use: Ornamental plant in the tropical areas. Often planted in parks or by riversides. The edible fruit is used to make wine. The root serves as a cure to the poison resulting from cassava.

Lecythidaceae

Barringtonia racemosa

English name: Fish poison tree, Sea poison tree

Taxonomic position: Genus Barringtonia, family Lecythidaceae

Description: Evergreen small tree with fissured barks. Branchlets are thick and stout, with distinct cicatricle; bracts a few to numerous at the base of apical buds; leaves leathery, usually apically clustered, short petiolate, entire; stipules small, caducous. Spikes are erect, with a cluster of bracts at the base of peduncle; both bract and bracteole caducous; floral buds globose; calyx tube obconical, lacerate or fissured at anthesis; calyx segments parallel veined; petals 4; stamens numerous; filaments folded in the bud; anther basal, usually fissured longitudinally in buds; style usually solitary.

Use: Ornamental plant. Fibre in the bark is used to make rope or wood for building. Root and fruit serve as medicine for fever and cough respectively.

● Iridaceae

Trimezia martinicensis

English name: Yellow walking iris, Forenoon yellow flag

Taxonomic position: Genus Trimezia, family Iridaceae

Description: Perennial herb. Leaves are ensiform, arising from basal rhizome; flowers flabellately arranged, small, golden, speckled with reddish brown stripes, blossoming from sheath-like bracts on the top of scape.

Use: Indoor potted plant, or often planted in the wetland or by riversides with strong adaptability.

Nepenthaceae

Nepenthes sp.

English name: Pitcher plant, Monkey cup

Taxonomic position: Genus Nepenthes, family Nepenthaceae

Description: Pitcher plant with a unique insect trap which absorbs nutrients from captured prey. Pitfall traps, modified leaves, are usually cylindrical, slightly ampliate at lower part, with hoods extending over their entrance. Inflorescence is raceme; flowers small, green or purple. Vase-shaped traps are a tool for capturing insects; ventral hoods (opercula) of the vase-shaped traps secrete smell to attract insects. The entrance surface of the trap is slippery, and insects slip into the trap and drown and die in the fluid on the bottom of the trap. Insects are digested by the fluids, and nutrients absorbed gradually by the pitcher plant.

Use: Ornamental potted plant.

Marantaceae

Ctenanthe setosa

Taxonomic position: Genus Ctenanthe, family Marantaceae

Description: Foliage is basal, long purple petiolate, green near base; leaf blades elliptical, silver, acute at apex, green along leaf veins, purple abaxially.

Use: Ornamental plant. Clustered planting in the shady place in parks, cool and refreshing.

Bignoniaceae

Pyrostegia venusta

English name: Flamevine, Orange trumpetvine

Taxonomic position: Genus Pyrostegia Presl, family Bignoniaceae

Description: Liana with 3-furcate filiform tendrils. Foliage is opposite; stamen borne in the central part of corolla tube; filament filiform; anther divaricate. Ovary is cylindrical, covered with dense pubescence; style thin; stigma linguiformly flattened. Both style and filament are projecting over corolla tube. Carpel is leathery, navicular; seeds multiseriate, winged, membranous. Flowers are red or orange red with a long flowering stage.

Use: Ornamental plant, climbing around pergolas. Or planted around buildings in courtyards or gardens.

Spathodea campanulata

English name: Fountain tree, Nandi flame

Taxonomic position: Genus Spathodea, family Bignoniaceae

Description: Evergreen tree, 10–20 m tall. Trunk is straight, grayish white, branched; leaves odd-pinnately compound; leaf blade elliptic or obovate, entire; leaflet short petiolate, oval-lanceolate or oblong; flower borne on the top of twig, about 10 cm long; corolla campanulate; tepal red or orange red.Flowers are densely clustered in a corymbose raceme looking like a burning flame. Capsule is oblong-quadrangular; carpel subligneous, reddish brown; seeds winged, membranous.

Use: Strong adaptability. Can be used as a hedge.

Pteridophyta

Huperziaceae

Phlegmariurus phlegmaria

English name: Coarse tassel fern, Common tassel fern

Taxonomic position: Genus Phlegmariurus Holub, family Huperziaceae

Description: Medium-sized epiphytic fern. Stems are caespitose; branches slender, pendulous, sulcate, 15–60 cm long; foliage leathery, triangular to lanceolate, heliciform, obliquely spreading, short petiolate. Strobilus is slender, pendulous, distinctly different from the infertile lower portion of fern, often multiple times dichotomously branched.

Use: Ornamental fern. Can be used as medicine.

● Lycopodiaceae

Lycopodium japonicum

Taxonomic position: Genus Lycopodium, family Lycopodiaceae

Description: Stems are repent; branches sparsely foliate Vegetative branches multiple times dichotomous, densely foliate; leaves aciculiform, with a caducous aristate long tail. Fertile branches are borne from foliage branches aged 2 or 3 years, much longer than the latter. Sporophyll is oval-triangular, acute at apex with a caudate tip, irregularly serrated at margin; Sporangia yellowish brown, reniform; spores reniform.

Use: Ornamental plant on woodsides or roadsides. Can be used as medicine.

● Aspidiaceae

Tectaria sp.

Taxonomic position: Genus Tectaria, family Aspidiaceae

Description: Rhizome is thick, creeping to upright, scaly at apex; scales lanceolate, dark brown; leaves fasciculate; leaf blades leathery or submembranous, triangular, unipinnate to tripinnate; sori usually rounded.

Use: Ornamental fern or potted plant. Also used for leaf-cutting.

Nephrolepidaceae

Nephrolepis cordifolia

English name: Fishbone fern, Tuberous sword fern

Taxonomic position: Genus Nephrolepis, family Nephrolepidaceae

Description: Rhizome is erect, covered with light brown, narrowly subulate scales; fronds fascicled, unipinnate or pinnately numerous, usually in a dense and imbricate arrangement; leaflets linear-lanceolate or narrowly lanceolate, alternate; leaf veins distinct; lateral veins slender; sori reniform, lining in a single row on both sides of midrib.

Use: Ornamental fern or potted plant. Also used as a leaf-cutting product. The tuber is edible and serves as medicine.

Nephrolepis hirsutula

Taxonomic position: Genus Nephrolepis, family Nephrolepidaceae

Description: Rhizome is short, upright, covered with blackish brown lanceolate scales; fronds are clustered, grayish brown, covered with scales, unipinnate, with 20–45 pairs of pinnae; leaflets or pinnae broad blanceolate or oblong-blanceolate, not overlapping one another, alternate, subsessile. Sori are rounded.

Use: Ornamental fern. Often planted in the shady place.

Nephrolepis acutifolia

Taxonomic position: Genus Nephrolepis, family Nephrolepidaceae

Description: Rhizome is short, upright, with blackish brown and subulate scales; fronds grayish brown, unipinnate, with 30–60 pairs of pinnae, covered with scales; pinnae lanceolate, alternate, far between pinnae, subsessile; sori rounded.

Use: Ornamental fern. Often planted in the shady place in parks or planted on trees.

Nephrolepis biserrata

Taxonomic position: Genus Nephrolepis, family Nephrolepidaceae

Description: Rhizome is short, upright, with reddish brown lanceolate scales; stolon borne from rhizome; fronds clustered, unipinnate; rachis sturdy, longitudinally grooved adaxially; pinnae narrowly elliptic, alternate, occasionally opposite, thin, papery, glabrous on both surfaces; sori rounded.

Use: Usually planted in gardens or courtyards for ornamentation.

● Aspleniaceae

Asplenium nidus

English name: Bird's-nest fern

Taxonomic position: Genus Asplenium, family Aspleniaceae

Description: Perennial shade herb, 80–100 cm tall. Rhizome is upright, short, thick, 2 cm in diameter, ligneous, dark brown, densely scaly at apex; scales broad lanceolate, about 1 cm long, acuminate at apex, entire, membranous, dark brown, slightly glossy.

Use: Potted plant. Also used for leaf-cutting. It helps to strengthen the physique, activate blood circulation and remove blood stasis.

Davalliaceae

Davallia mariesii

English name: Squirrel's foot fern
Taxonomic position: Genus Davallia Sm., family Davalliaceae
Description: Epiphytic fern. Rhizome is long-creeping; scales broad lanceolate or lanceolate; fronds arising from long aerial rhizome, quadripinnate; rachis dark straw-colored or brown; pinnae pentangular, opposite, stipitate, obliquely spreading; lobes elliptic; sori borne at apex of veinlets; indusium tubulose, truncate at apex, brown, thickly membranous.

Use: Ornamental potted plant in courtyards. The rhizome helps to tonify kidney and strengthen bones.

Davallia divaricata

Taxonomic position: Genus Davallia Sm., family Davalliaceae

Description: Epiphytic fern. Rhizome is thick, stout, creeping; both rhizome and basal petioles densely covered with bright brown scales; scales lanceolate, slightly serrated; fronds deltoid, acuminate and pinnate at apex, quadripinnate or quinpinnate below apex; sori numerous, usually borne on the base of upper forked veinlets, with 1 sorus on each tooth of last-pinnate lobe; indusium cup-shaped, truncate at apex, with golden lustre.

Use: Ornamental fern in courtyards. The rhizome helps to tonify kidney and strengthen bones.

Vittariaceae

Vittaria ophiopogonoides

Taxonomic position: Genus Vittaria, family Vittariaceae

Description: Rhizome is creeping, densely scaly; scales yellowish brown, glossy, subulate-lanceolate; petioles short, slender, pale brown at lower part, covered with slender small scales; leaves linear, thin, leathery, revolute on edge covering sori; sori linear, borne on the inner side of margin in shallow grooves; spore elongate elliptical, transparent, colorless, monolet, adaxially with blurry granular stripes.

Use: Ornamental fern in the tropical areas.

Vittaria taeniophylla

Taxonomic position: Vittaria, Vittariaceae

Description: Rhizome is short, creeping; scales pale brown, subulate; leaves fasciculate, leathery, sessile; sori mostly borne on upper surface or in slightly concave grooves.

Use: Ornamental fern in the tropical areas.

Polypodiaceae

Microsorium punctatum

English name: Fishtail fern

Taxonomic position: Genus Microsorum, family Polypodiaceae

Description: Rhizome is short, thick, stout, creeping, with a few annular bundle sheaths, mostly with sporadic sclerenchyma, subglabrous, covered with white powder, densely rooted with fibrous roots, sparsely scaly; foliage nearly fasciculate; short thick petiolate or subsessile, broad linear-lanceolate, acuminate at apex, attenuate to narrow wing at base; sori orange, generally borne at apex with included veinlets; spores phaseoliform, with flat to tuberculiform perisporia.

Use: Ornamental plant. Also used as medicine.